Mathematical Cartoons

Ideas by

Charles Ashbacher
5530 Kacena Ave
Marion, IA 52302 USA

cashbacher@yahoo.com

Drawn by

Caytie Ribble

Einsten image by

Jenna Richardson

ISBN-13: 978-1514207130

CONTENTS

Author's Note

 This book in one that I have wanted to develop for years. I am a big fan of the intelligent cartoon, one that not everyone will immediately understand when they see it. In my years as co-editor of **Journal of Recreational Mathematics** I occasionally asked for people to submit mathematical humor, but only rarely did I receive anything in that area. Therefore, I realized that the only way I could publish mathematical humor was if I drove the creation myself.

 Each of the cartoons in this collection was drawn by Caytie Ribble and based on my ideas. The mathematics covers a broad spectrum and there are a few references to popular culture as well. For example, the first three reference the Star Trek™ original series. The second section of the book contains a brief explanation of all of the cartoons.

The last image in this collection is one of Albert Einstein and was drawn by Jenna Richardson.

 As always, I am very interested in hearing from readers regarding their opinion of the content of this book.

Charles Ashbacher

cashbacher@yahoo.com

HOW IT STARTED

LANDRU ALGEBRA

$$A \cup (A \cap B) = A$$

$$A \cap (A \cup B) = A$$

The Lights of Zeta-R

ANTECEDENT

$$\int f(x)\,dx = -C + F(x)$$

A posteriori reasoning

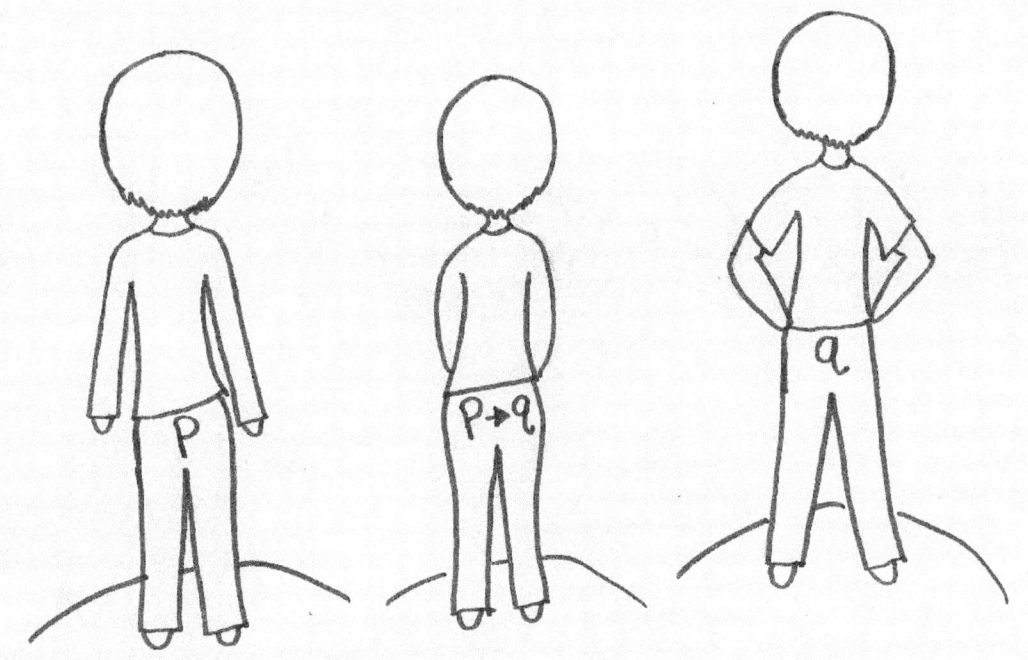

CATENARY CURVE

CIRCULANT

7 58 24 1

4 1 7

2 1/7 5

8 8

5 7.14 2

POOF BY DESCARTES

devil's curve

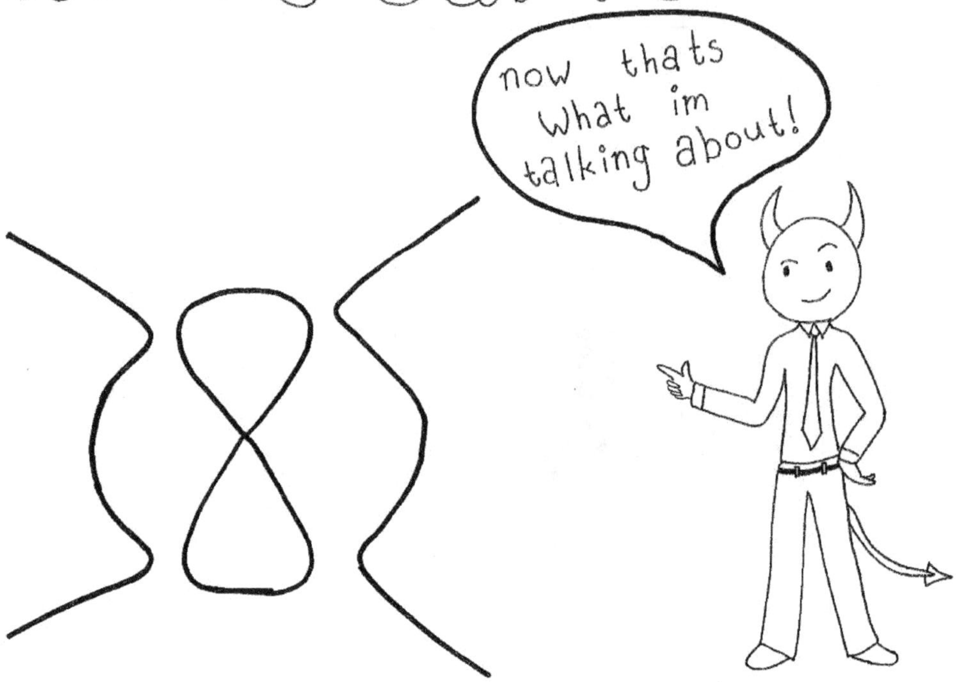

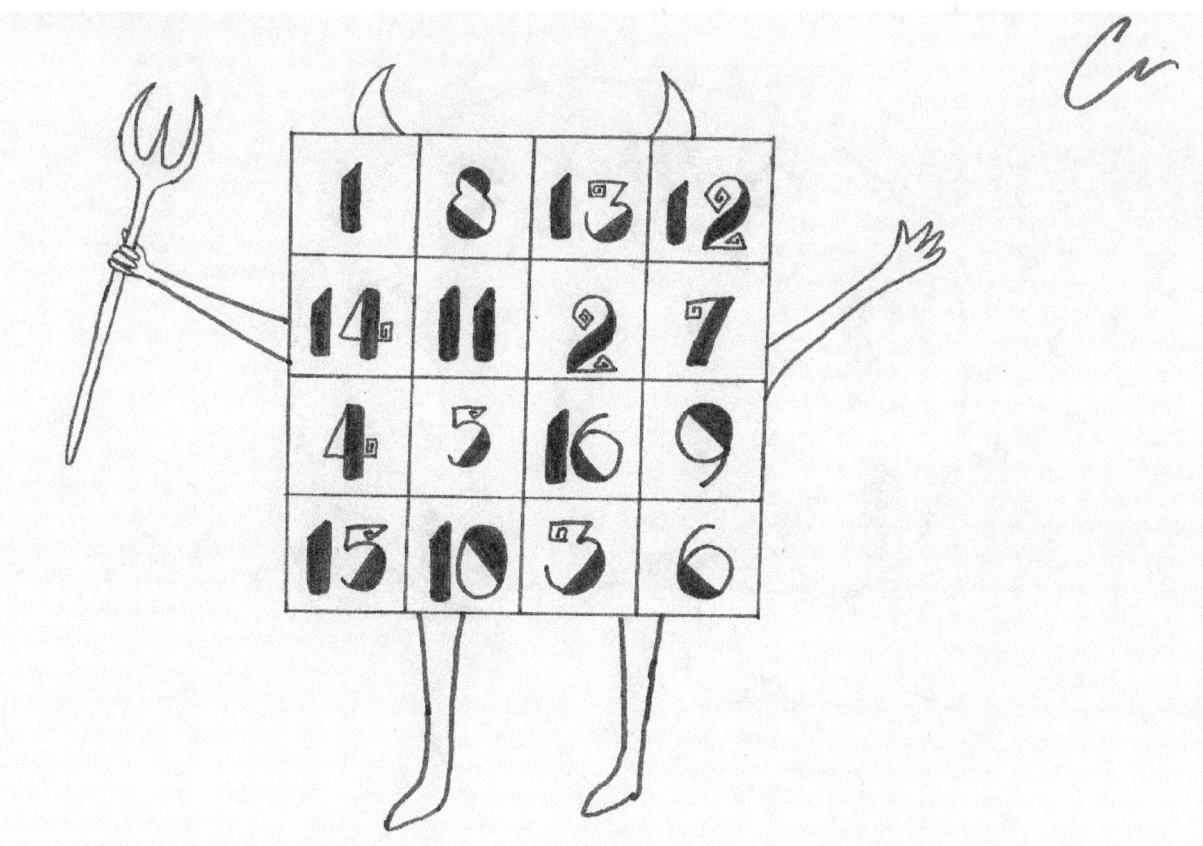

DIABOLIC MAGIC SQUARE

DIEAGRAM

DIELATION

DIEGRAPH

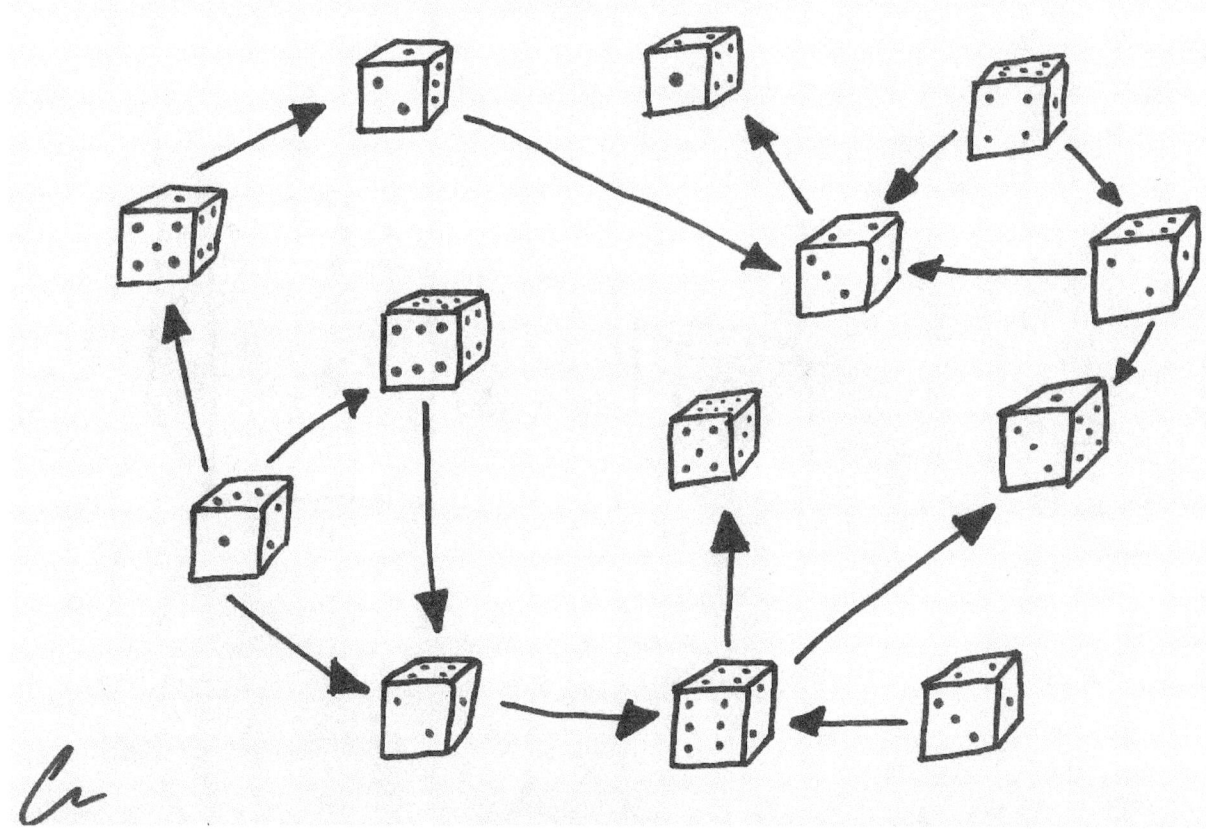

DUODECIMAL

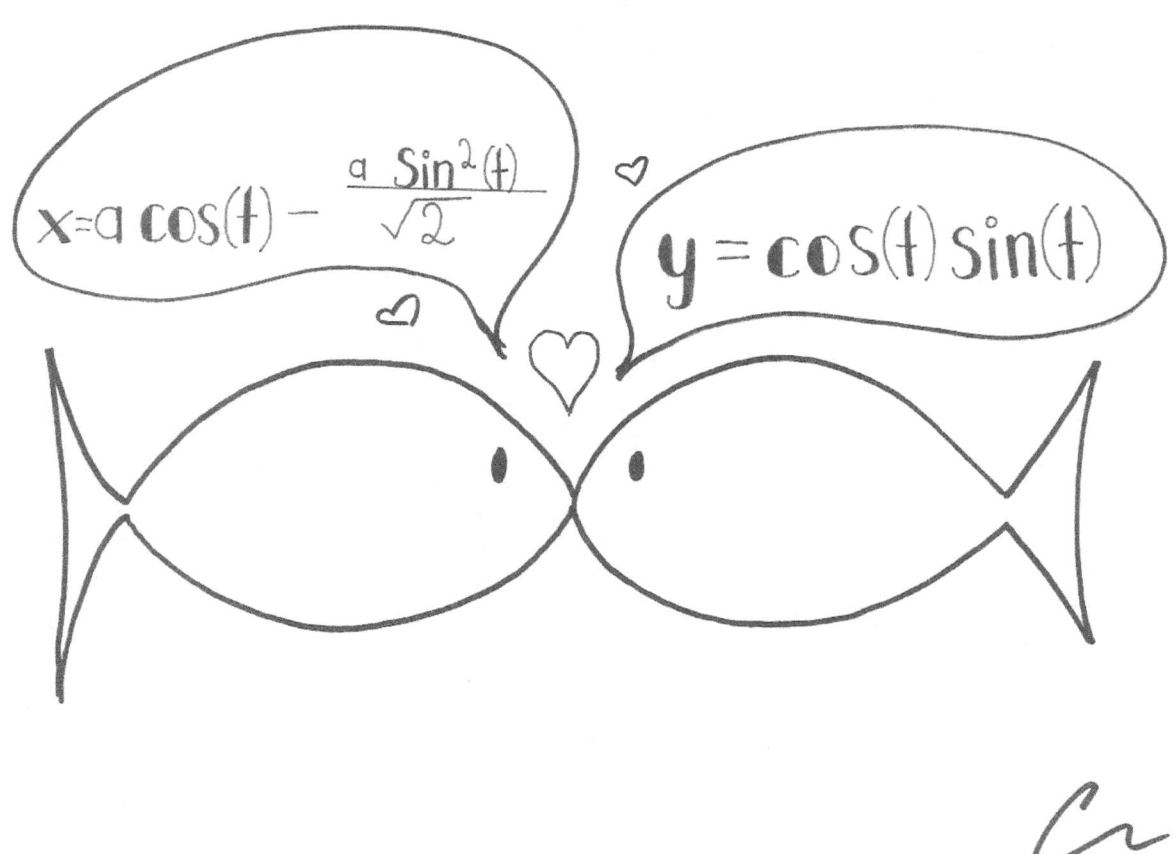

HIPPOPEDE CURVE

HYPERCUBE

ICE-OMORPHISM

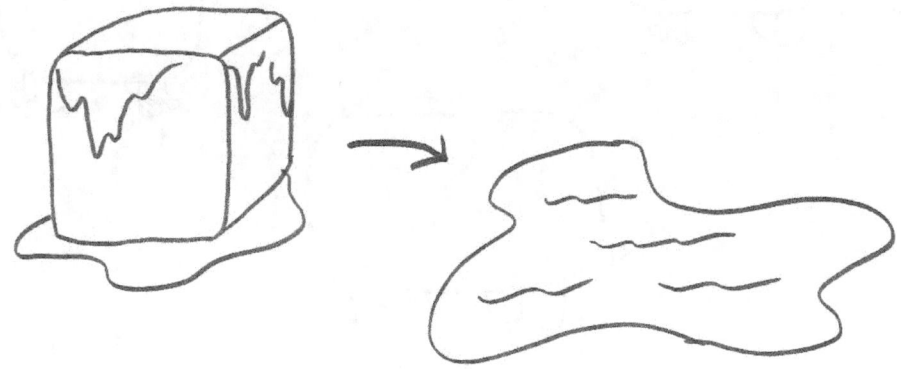

LAMÉ CURVES

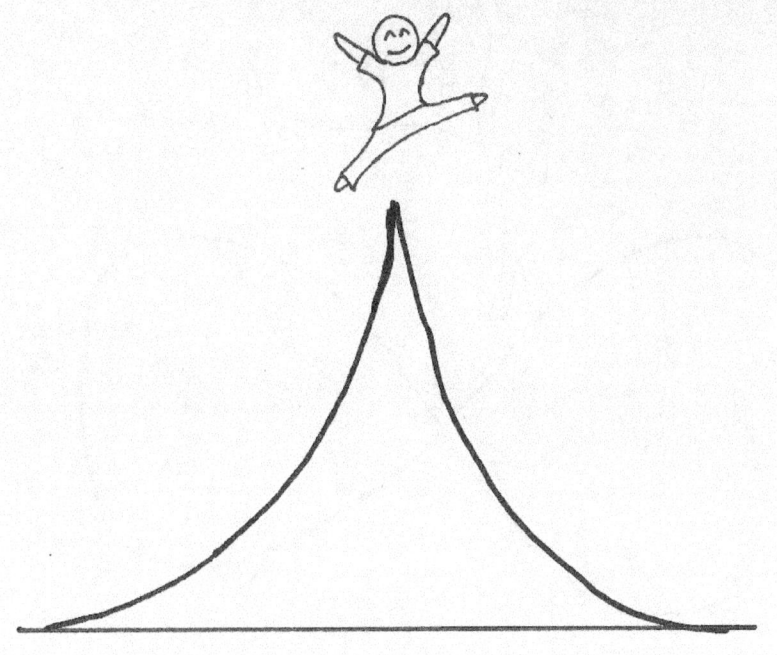

LEAPTOKURTOSIS

Lemniskate

A DOUBLY-TRUE INTEGRATION

$$\int \frac{d \ ical}{ical} = \log \ ical$$

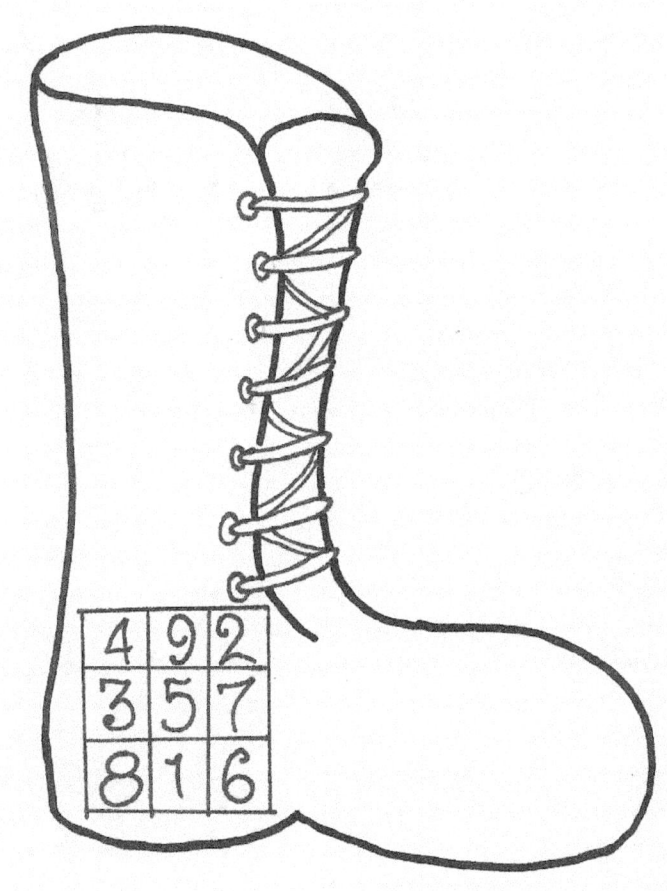

Luoshoe

HOW A MATHEMATICIAN SNORES

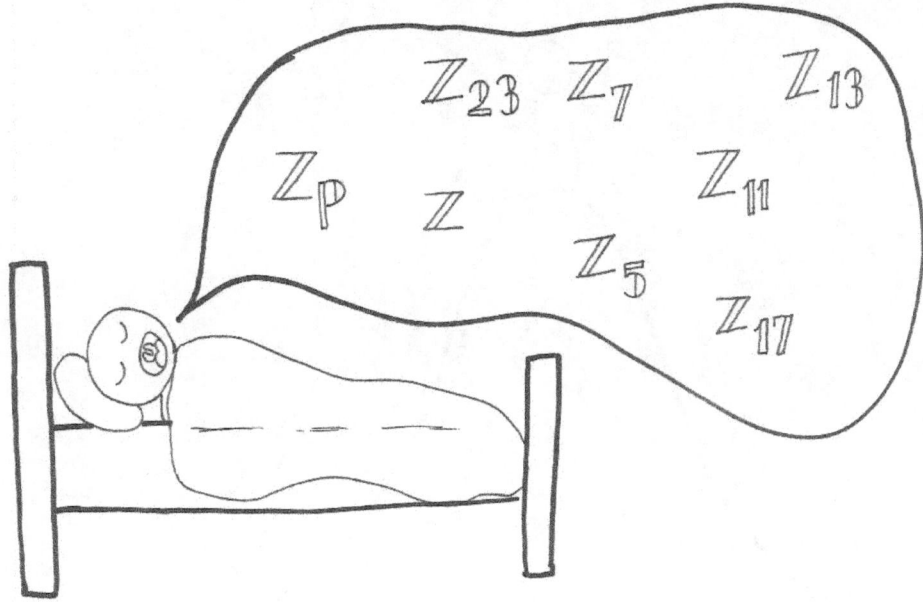

A normal curve is perfectly balanced about the mean

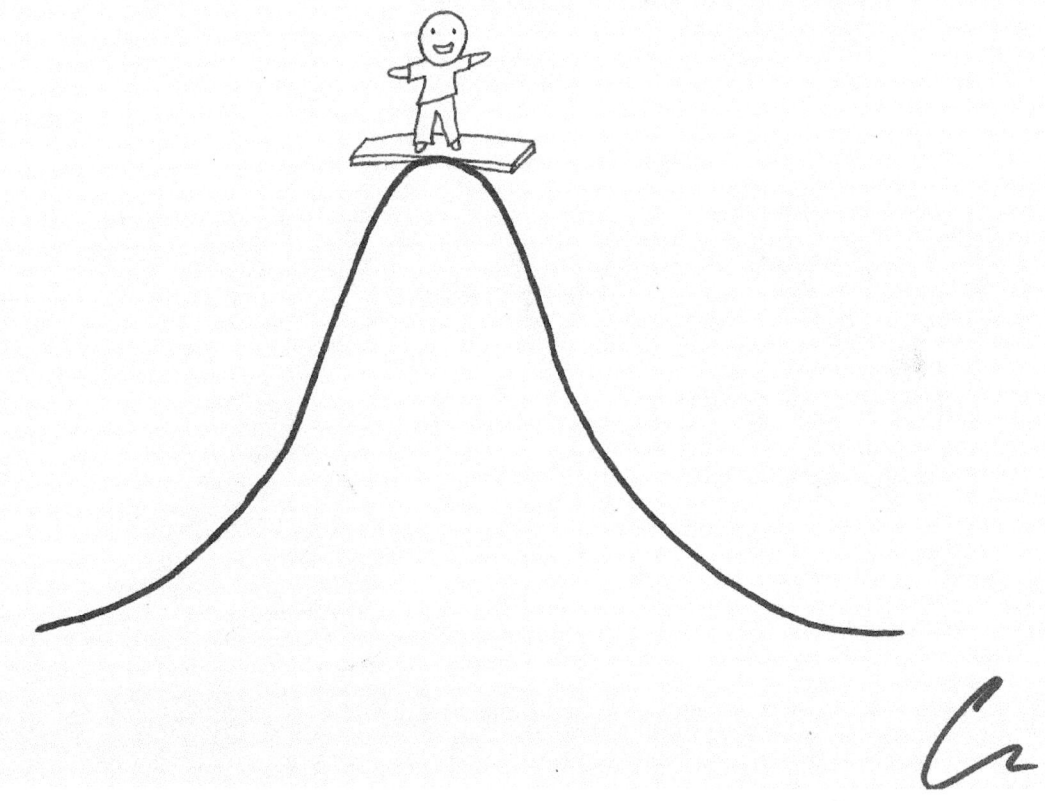

OPERAND

platypotic curve

SERPENTINE CURVE

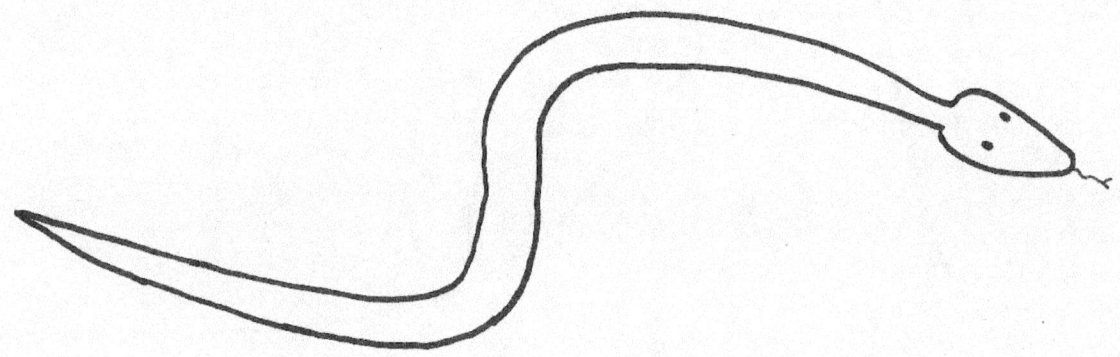

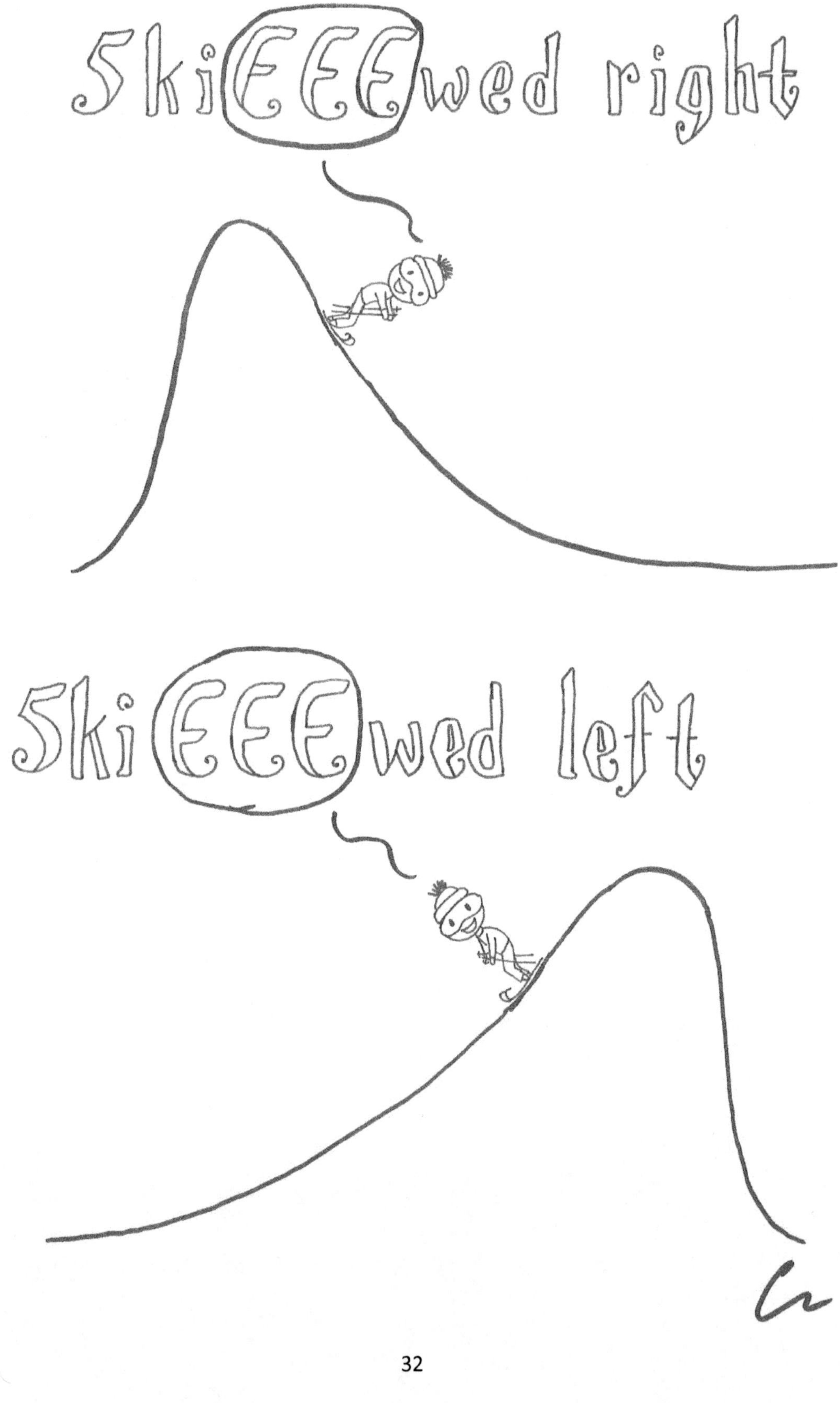
SkiEEEwed right
SkiEEEwed left

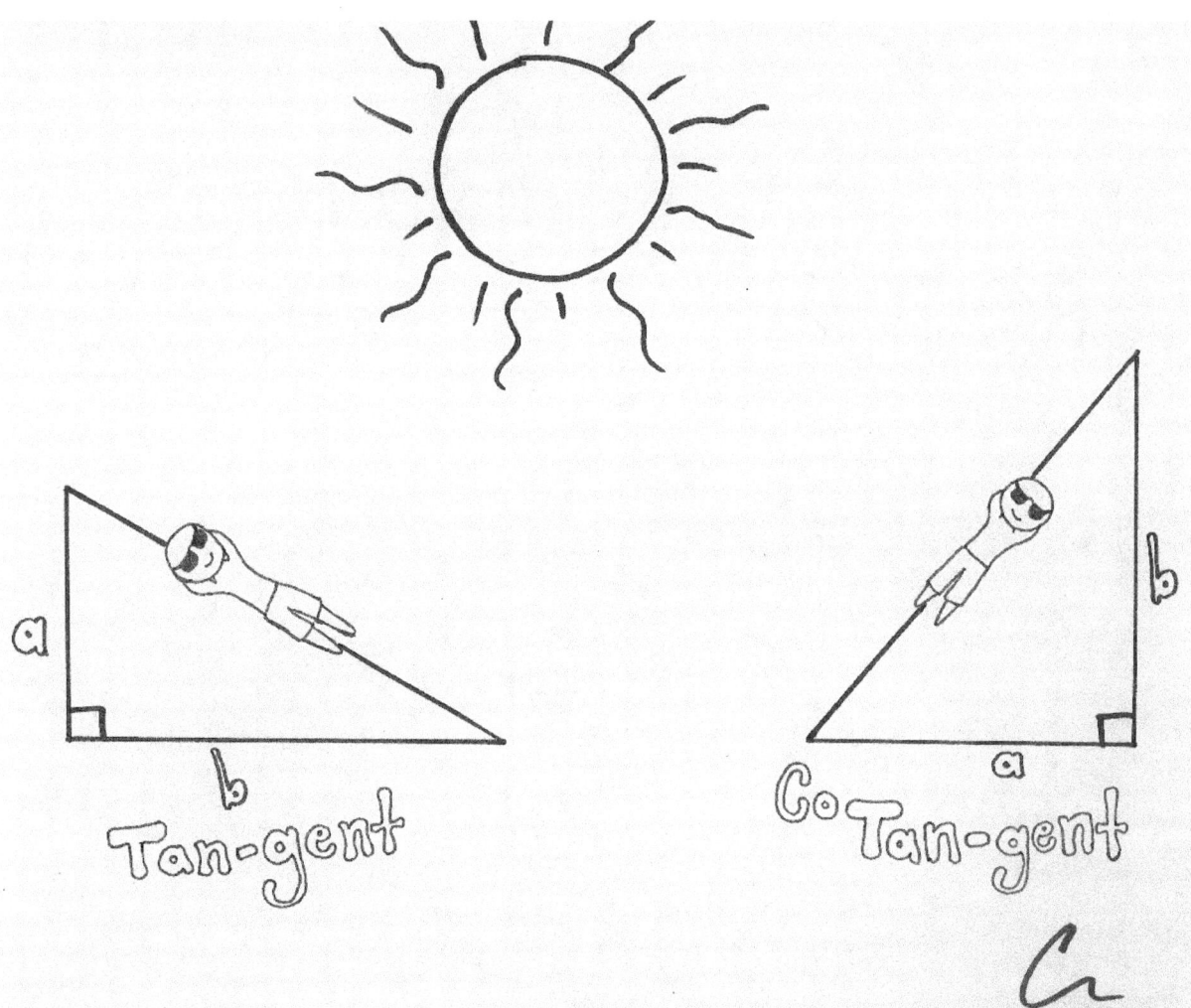

Thoughts of a Setistic Person

Tractrix

TRiMONSTER

$2^{46} \cdot 3^{20} \cdot 5^{9} \cdot 7^{6} \cdot 11^{2} \cdot 13^{3} \cdot 17 \cdot 19 \cdot 23 \cdot 31 \cdot 41 \cdot 47 \cdot 59 \cdot 71$

$2^{46} \cdot 3^{20} \cdot 5^{9} \cdot 7^{6} \cdot 11^{2} \cdot 13^{3} \cdot 17 \cdot 19 \cdot 23 \cdot 29 \cdot 31 \cdot 41 \cdot 47 \cdot 59 \cdot 71$

$2^{46} \cdot 3^{20} \cdot 5^{9} \cdot 7^{6} \cdot 11^{2} \cdot 13^{3} \cdot 17 \cdot 19 \cdot 23 \cdot 31 \cdot 41 \cdot 47 \cdot 59 \cdot 71$

UNICURSAL

VERSED SINE

If Galileo had said in verse That the world moved the Inquisition might have Left him alone

Vulgar Fractions

Waltzing
The
Tilde

Yotta-Yotta-Yotta

$$10^{24} \cdot 10^{24} \cdot 10^{24}$$

NDOMORPHISM

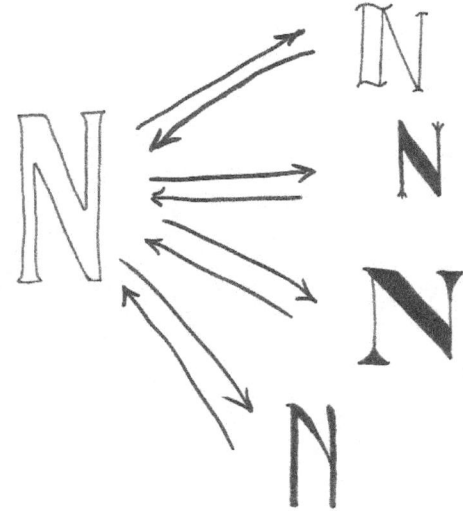

GREEK SHEEP

Wearing a Cocked Hat

$$y^2(a^2 - x^2) = (x^2 + 2ay - a^2)^2$$

THE PROPER WRAP TO
A MATH PRESENT

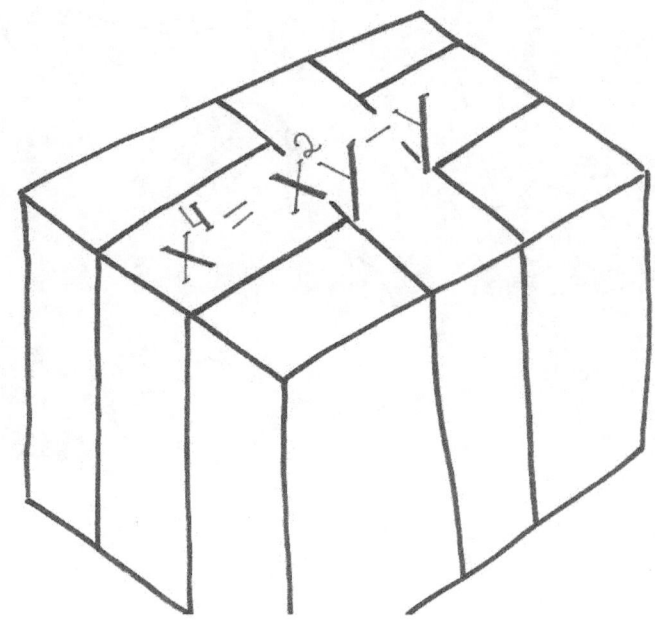

MATHEMATICAL COWS

mathematical trident

$xy + ax^3 + bx^2 + cx = d$

$xy + ax^3 + bx^2 + cx = d$

$xy + ax^3 + bx^2 + cx = d$

$xy + ax^3 + bx^2 + cx = d$

$xy + ax^3 + bx^2 + cx = d$

$xy + ax^3 + bx^2 + cx = d$

xy

monkey Saddle

$$z = x(x^2 - 3y^2)$$

COMMON ERROR MADE BY TOPOLOGISTS

Ten – or

MATHEMATICAL COXWAIN

Explanations of the cartoons

How it Started – the second pilot of the Star Trek original series was called "Where No Man Has Gone Before" and an edited version was the third episode of the original series and given the same title. The cartoon depicts the bridge of the Enterprise with an alphametic displayed on the view screen.

Landru Algebra – in the mathematical construct known as Boolean Algebra, the two formulas are known as the absorption laws. In the Star Trek original series episode "The Return of the Archons" the crew of the Enterprise encounter a planet where the people are absorbed into a collective consciousness known as "the body." The leader of the body is a computer and has the title "Landru."

The Lights of Zeta-R - in the Star Trek original series episode, "The Lights of Zetar" the Enterprise encounters a set of incorporeal creatures that are represented as lights and claim to be from the planet Zetar. The summation on the left is the Riemann Zeta function that has been set equal to R.

Antecedent – This is a double pun. The result of taking an antiderivative is generally expressed as

$$F(x) + C$$

where C represents an arbitrary constant. By putting the C in front and making it negative it is both the antecedent as in coming before, as well as an "anti-C." Since C is an arbitrary constant, the mathematical expression also remains valid.

A Posteriori Reasoning – in logic, the sequence $p \rightarrow q$, p then infer q is called "modus ponens." A posteriori knowledge is based on experience of empirical evidence. The location of the logic symbols provides the pun.

Catenary Curve – this is the shape formed by a hanging chain or a necklace and is generated by the expression

$$y = a * \cosh(x / a)$$

where cosh is the hyperbolic cosine function. The creature on the curve is of course part cat and part canary.

Circulant – the fraction 1 / 7 has the repeating decimal value $0.\overline{142857}$, in this cartoon a circle is formed that repeats this indefinitely. The term "circulant" refers to a matrix where the rows offset-repeat a sequence of numbers.

Poof By Descartes – mathematician/philosopher René Descartes famously said, "I think, therefore I am." In this case if he thinks not he is not.

Devil's Curve – The curve seen in the cartoon is called a Devil's Curve and they are defined by the general equation

$$y^2(y^2 - a^2) = x^2(x^2 - b^2).$$

The commentator is of course a devil.

Diabolic Magic Square – when the diagonals of a magic square also sum to the the magic sum, the square is called "pandiagonal" or "diabolic." It comes from the prefix "dia-," which means through or across. It is a pun, and since the Greek word "diabolos" also means devil, the accoutrements were added.

Dieagram – this is a simple pun on the word "diagram," the die being weighed is exactly one gram.

Dielation – a dilation is a process where an object is magnified, while retaining the same general appearance. In this cartoon a pun is created by having the object be a die.

Diegraph – a directed graph or digraph is a construct of nodes and directed connections. In this case a pun is created by having the nodes represented by dice.

Duodecimal – the base 12 system is called duodecimal and in this case the duo together makes 12.

Fish curves – the equations in the dialog balloons are the parametric equations for a curve known as the "fish curve" due to its shape.

Hippopede curve – the equation $(x^2 + y^2)^2 + 12(x^2 + y^2) = 36x^2$ has a graph that looks like the curve in the figure and is given the name hippopede. The face is of course that of a hippopotamus.

Hypercube – a hypercube is a cube in four or more dimensions. The prefix "hyper-" also means nervous and this cube is nervous.

Ice-omorphism – an isomorphism is a correspondence between two sets that is 1-1 and has an interpretation where they are essentially the same. The molecules of an ice cube and the resulting liquid could be placed in a 1-1 correspondence.

Lame curves – a Lamé curve is also known as a superellipse and is defined by the equation

$$\left| \frac{x}{a} \right| + \left| \frac{y}{b} \right| = 1.$$

The bandaged legs makes this a pun on the word "lame."

Leaptokurtosis – a leptokurtic curve is one heavily concentrated about the mean. In this case that characteristic is exaggerated to a point and the person is taking a daring leap over the kurtotic feature.

Lemniskate - the lemniscate curve is a figure eight that is flat and is also the symbol for infinity. This is a pun on the term "skate" rather than "scate."

A Doubly-True Integration – using the variable name "ical" this expression is both correct as an integration and sends a message.

Luoshoe – the Lo Shu is the magic square seen in the image and has its origins in Chinese folklore. This is a pun on the word "shoe."

How a mathematician snores – in this case the Z's representing sleep are all mathematical expressions where Z is used to represent a set of integers.

Normal curve – since the normal curve is perfectly balanced, the mean point is the only location where one could balance like that.

Operand – the expression $p \wedge q$ is logical "and" and the woman is an opera singer. The $\wedge$ is also an operator, increasing the level of the pun.

Platyputic curve – a platykurtic curve has the shape seen in the image. The animal is a platypus, making the complete pun.

Serpentine curve – the serpentine curve is one that has a shape similar to the snake in the image, which makes the pun. Note the snakes in the lettering.

Skieeewed – a skewed curve is one where the data is concentrated to one side. In this case the person is skiing on the side that is skewed.

Super Power – the expression x^x is known as the superpower, which is the emblem on the chest of this female superhero.

Tan-gent – the **gent**leman is getting a tan and his slope is a/b. His co-tanner has a slope of b/a, consistent with the definitions of the tangent and cotangent.

Thoughts of a setistic person – the thoughts are all of items from set theory and this is a pun on the word "sadistic."

Tractrix – the curve in the image is called a tractrix and the person is performing a trick by balancing on one foot on the point.

Trimonster – the number $2^{46} * 3^{20} * 5^9 * 7^6 * 11^2 * 13^3 * 17 * 19 * 23 * 31 * 41 * 47 * 59 * 71$ is the number of elements in the largest sporadic finite simple group. It is called the monster group for obvious reasons.

Unicursal – a unicursal figure is one that can be drawn by putting a pencil down and never having to lift it up. The unit here is also cursing and the unit figure is itself unicursal.

Happy Valentine's Day – when plotted, the formulas on the card will generate a heart-shaped cardioid.

Versed sine – the versed sine is defined as $\sin\theta = 1 - \cos\theta$. In this case, the pun is the short section of verse on the sine curve.

Vulgar fractions – in old terminology a vulgar fraction was one that was in lowest terms. In this case the pun is that they are being verbally vulgar.

Waltzing the tilde – in Australia, the phrase "waltzing Matilda" refers to the act of traveling on foot with one's belongings in a bag or Matilda on their back. In this case, the pun is the mathematical symbol tilde ($\sim$) on the backpack.

Yotta – yotta – yotta – the phrase "yada yada yada" was used in the "Seinfeld" sitcom and is used as an equivalent to "and so on" or "blah blah blah." In mathematics, yotta is the prefix meaning 10^{24}.

Ndomorphism – in mathematics, an endomorphism is a mapping that maps a set onto itself, which is what each of these arrow transformations are doing.

Greek sheep – the character Λ is the uppercase lambda so the animals are all "lambdas."

Wearing a cocked hat – when plotted, the formula $y^2(a^2 - x^2) = (x^2 + 2ay - a^2)^2$ generates an image that resembles a bicorne or cocked hat. Images of Napoleon depict him as wearing such a hat.

The proper wrap to a math present – when plotted, the formula $x^4 = x^2y - y$ generates an image that looks like a bowtie.

Mathematical cows – these cows are uttering the Greek letter "mu", commonly used to represent the mean, rather than the standard "moo."

Mathematical trident – the plot of the formula $xy + ax^3 + bx^2 + cx = d$ is known as the trident curve.

Monkey saddle – the plot of the formula $z = x(x^2 - 3y^2)$ is known as the monkey saddle.

Common Error Made by Topologists – in the classification of objects according to their topological properties, a doughnut is equivalent to a coffee mug with a fingerhole. Therefore, the joke is that a topologist cannot tell the difference between a doughnut and a coffee cup.

Ten-or – the mathematical operation of $p \lor q$ is known as logical "or." The person is a male opera singer or tenor, making the pun.

Mathematical coxwain – the person in the stern of the boat is called the coxwain (also spelled coxswain). It is commonly believed that they count out the cadence "stroke – stroke – stroke" to keep the rowers synchronized. The up arrow $\uparrow$ is a symbol in logic known as the "Sheffer stroke" and behaves as the "not-and" connective, making the pun.

BOOKS BY CHARLES ASHBACHER AND ASSOCIATES

Topics in Recreational Mathematics 1/2015 ISBN 978-1507603215

Topics in Recreational Mathematics 2/2015 ISBN 978-1508617099

Topics in Recreational Mathematics 3/2015 ISBN 978-1511641005

Alphametics as Expressed in Recreational Mathematics Magazine ISBN 978-1508538134

Ten Year Cumulative Index to the Journal of Recreational Mathematics, edited by Joseph S. Madachy and Charles Ashbacher ISBN 978-1508936800

Alphametics Expressing Thoughts From the Star Trek Original Series ISBN 978-1512152784

Associates

Artist Catie Ribble

Editor Rachel Pollari

Editor Jennifer Corrigan

Artist Jenna Richardson

www.ingramcontent.com/pod-product-compliance
Lightning Source LLC
Chambersburg PA
CBHW080607180526
45168CB00007B/2814